The Spiders Dance

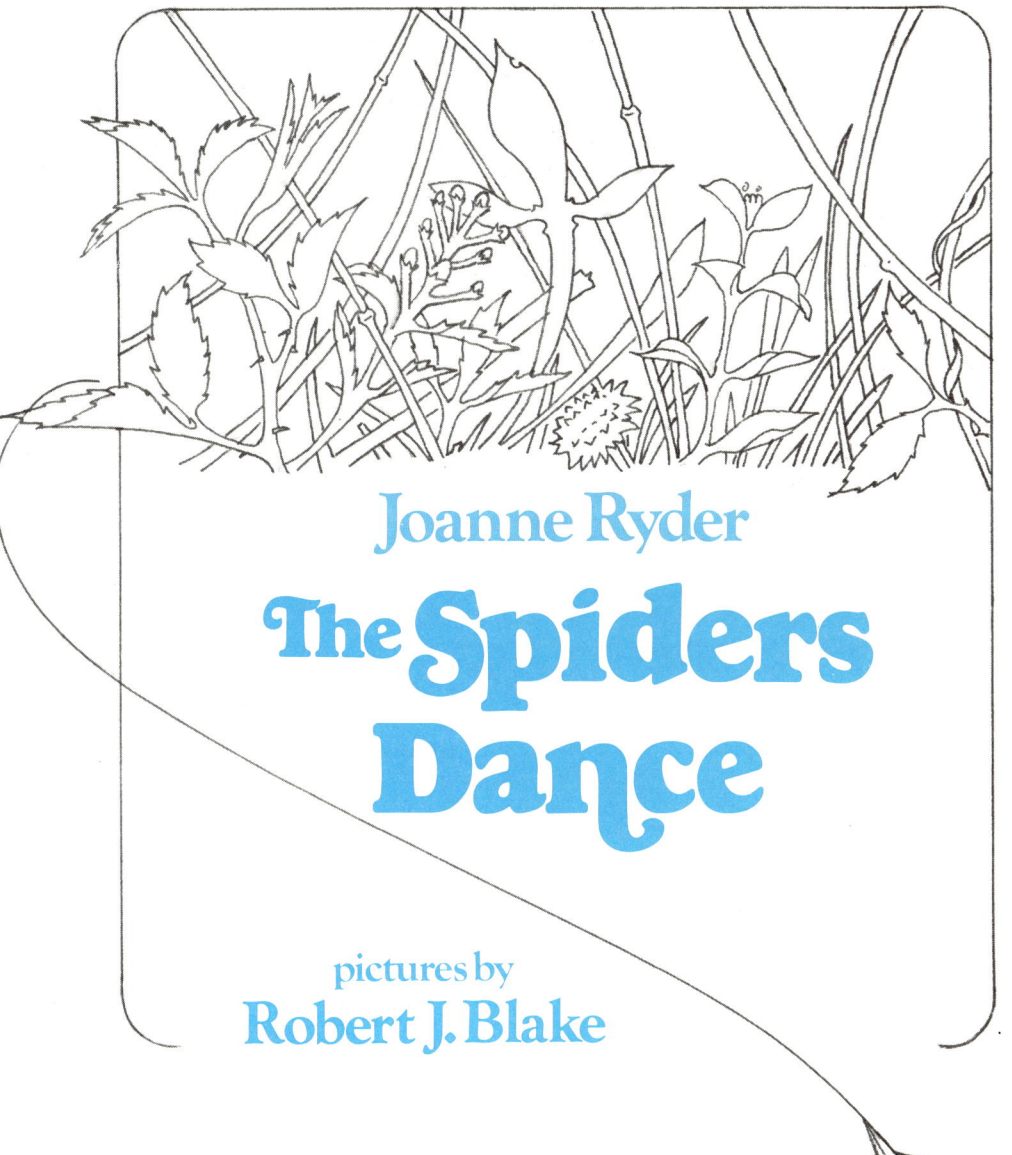

Joanne Ryder

The Spiders Dance

pictures by
Robert J. Blake

Harper & Row, Publishers

The Spiders Dance
Text copyright © 1981 by Joanne Ryder
Illustrations copyright © 1981 by Robert J. Blake
All rights reserved. No part of this book may be
used or reproduced in any manner whatsoever without
written permission except in the case of brief quotations
embodied in critical articles and reviews. Printed in
the United States of America. For information address
Harper & Row, Publishers, Inc., 10 East 53rd Street,
New York, N.Y. 10022. Published simultaneously in
Canada by Fitzhenry & Whiteside Limited, Toronto.
FIRST EDITION

Library of Congress Cataloging in Publication Data
Ryder, Joanne.
   The spiders dance.

   SUMMARY: When tiny spiders burst out of their eggs,
the dance of the spider's life cycle begins.
   1. Spiders—Juvenile literature. [1. Spiders]
I. Blake, Robert J. II. Title.
QL458.4.R92 1981   595.4'4   78-22495
ISBN 0-06-025133-6
ISBN 0-06-025134-4 (lib. bdg.)

To my cousins

    Elizabeth
    Gillian
    Ingrid
    Kerin
    Shannon

The Spiders Dance

In the bushes
a silken ball bursts open.
Tiny spiders scramble
    over tiny spiders
        over tiny spiders.
They stretch each new long leg,
so many legs,
and dance along the branches of the bush.

pit patter pit pit patter
Spring rain dribbles
from leaf
       to leaf
             to leaf.

The tiny spiders huddle
in a ball of moving legs—
until the sun calls one
to lead the others
ever upward.

Up a lean green tower
the leader climbs,
to its leafy top.

She spreads her legs
to form a circle
and spins the thinnest thread.

Soon other spiders
spin fine lines
to catch the wind.
And *pooooofff*
                let go.

On silken streamers
tiny spiders fly
    past tiny spiders
        past tiny spiders
and leave a silver bush behind.

The tiny spiders ride
the teasing breeze
this way and that way
over ponds and fields and shady woods.

They float above the homes of other spiders—

past speckled spiders hiding under stones,
past small gray spiders leaping
through the grass like frogs,
past lean swift spiders
running over piles of brown wet leaves,

and past a yellow spider in a yellow flower
never moving
till a foolish fly swoops by
too close, too close!

The tiny spiders fly past houses
where inside
spiders in the cellar
spin loose light webs
around the half-used cans of paint,

and spiders in the attic
weave patterns 'round grandmother's clothes
and under one small bed
a small black spider
spins a messy web.

Outside, the tiny spiders fly
above all sorts of places
with all sorts of spiders in them.

Reeling in their lines,
the spiders twirl,
floating down
to places of their own.
Spiders rain
here
      and there
            and over there.

Drifting down and down,
the leader touches grass.
Her small round body glides high
on four legs east and four legs west
as she explores her field.
She picks *her* place, her home,
and climbs inside her bush.

One evening
as the sun sets low,
the spider spins a bridge
across an empty space.
She builds
a fragile crooked wheel
with many silky spokes.
Softly, the wind whispers
through her round full web,
silver in moonlight.

She waits frozen in slow spider time
until the thin moon hides
behind the clouds.

In the dark
a small light flashes bright.
Soft sounds of wings.
A firefly sails through the branches
into the space—
no longer empty.

The web shivers.
The spider dances swiftly,
trapping the firefly
and wrapping him
in a tight tangle of silk,
until the glowing firefly
fades
into blackness.

Holding him close,
the tiny spider dances
from her torn and tattered web.
Tucked tight,
she eats her meal
inside her leafy nest.

In the morning light
a cloud of flying aphids
drifts through the bush,
calling the spider once more.

Later she rolls
the fragments of her web
into a tangled ball
and eats that too.
And ever circling,
she dances in the air
to weave another web.

Snipping, snapping,
black bird's beak
snips too close.
The spider hides,
but not quite fast enough.
On seven legs,
her jig is awkward—
yet she spins.

Twitching, twisting, twirling,
the swollen spider
kicks and kicks her way
out of her skin.
Each time she grows
she sheds her skin
for larger, hairy suits of clothes
and dances on her new bright legs—
eight legs once more!

Dangling,
her old skin
haunts the web.

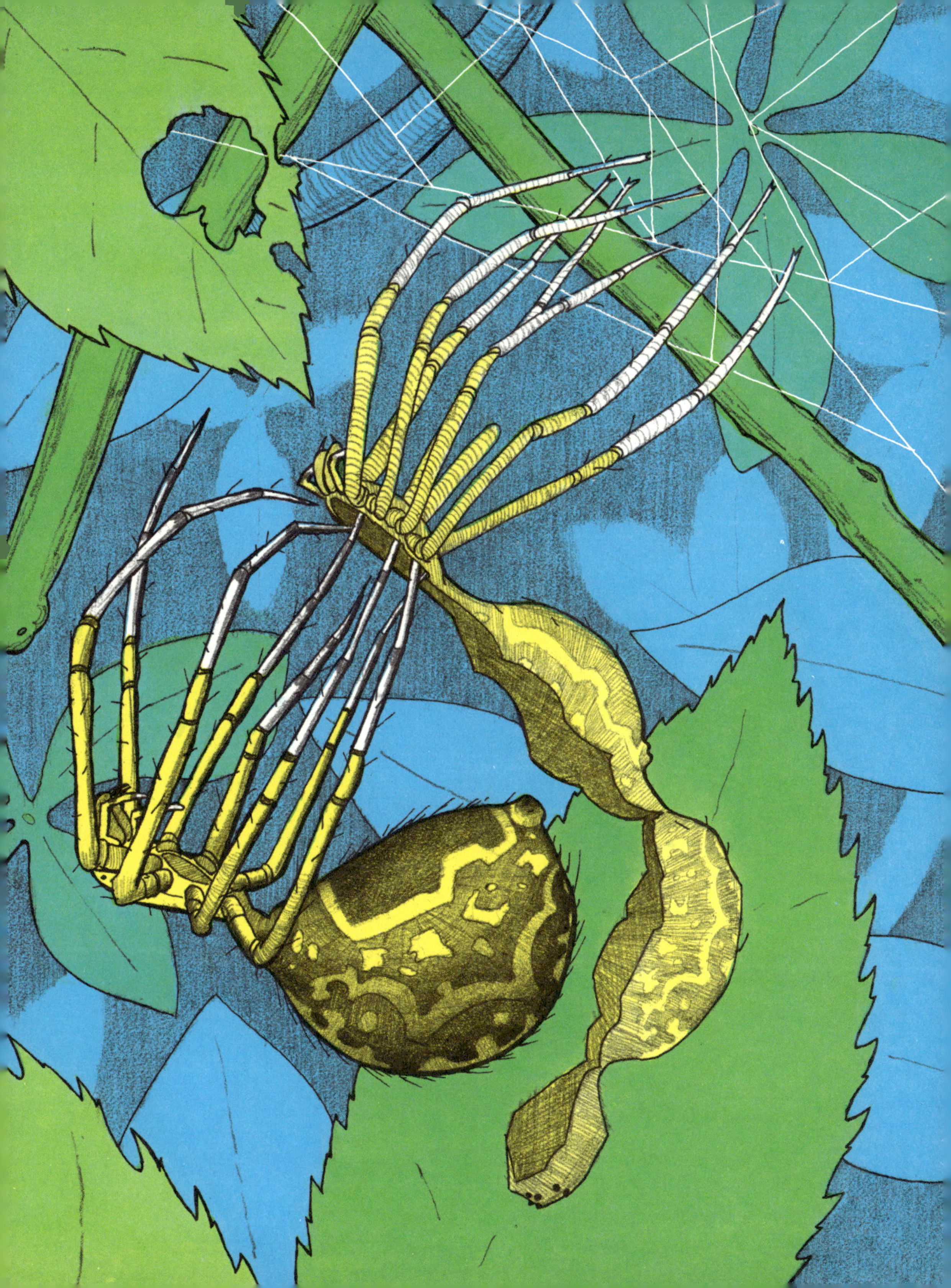

All summer long
the growing spider hunts.
Her big, bright body
leaps across her larger webs.

The black bird
sinks into the bush,
hunting with a cold, gold eye.
He sees the spider!
She drops
a thousand spider heights
into the grass.
He paces through the greenness
searching,
finding no one.

But when he flutters off,
a hidden pebble moves,
kicks, turns over,
and climbs up
her long long stairway home.

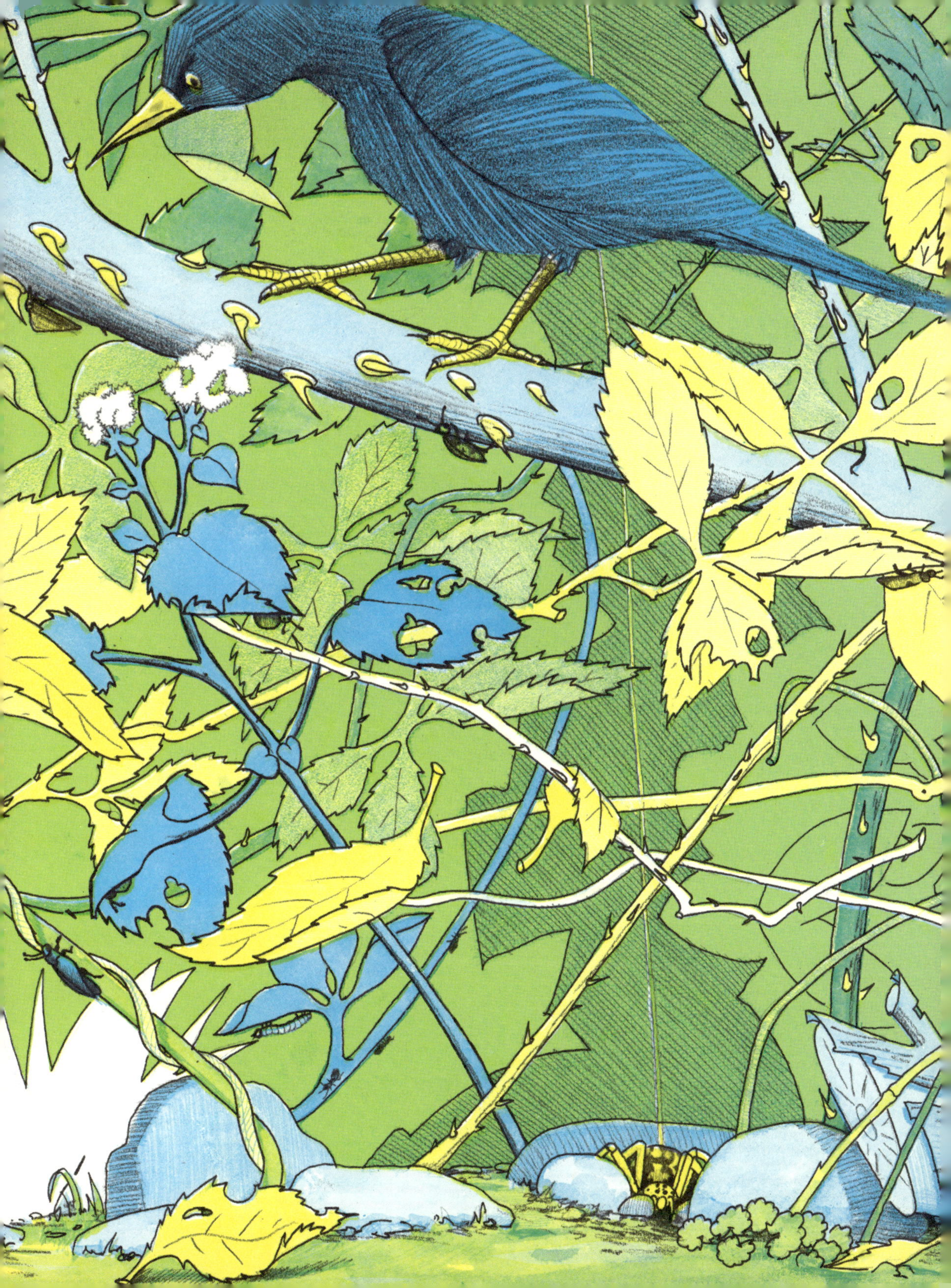

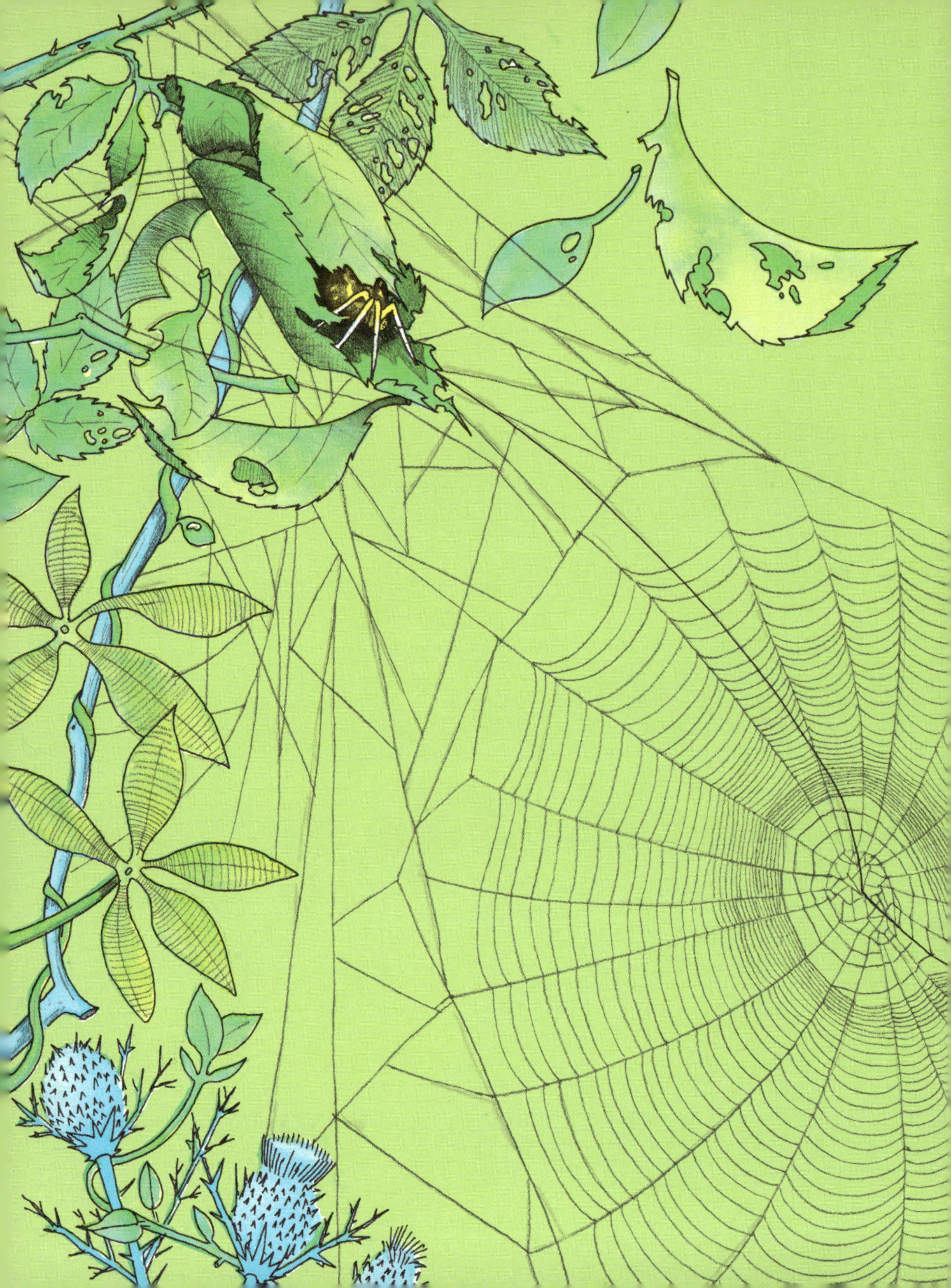

Ten, twenty,
thirty, forty,
fifty, sixty webs
the spider spins
before the days grow cold
and all the leaves around her home
turn red and gold.
Time is changing everything.

Until, one day, her web begins to tremble—
softly, gently.
Someone is near,
tugging at her home.
The spider waits
and answers with a tug
until the stranger glides to her.
At last, two spiders
dancing their own dance
inside the web.

One cool morning
someone small and careless
flies into the web,
but no one dances down
to find him there.
The leaf above is empty.

Deep inside another bush,
the spider creeps,
fat and full,
looking once more
for a special place.
Underneath a branch,
she spins a sheet of silk
and lays her eggs—
more than one hundred eggs—
inside a yellow fluffy ball.

She hides them
from the wind
and from the rain
and from the black birds,
spinning and spinning
until she is too tired
to ever spin again.

So very tired,
she bows
and leaves her eggs behind.

In wintertime
the cold bare bush
displays the spider's gift,
a yellow ornament
full of life,
waiting for spring
when tiny spiderlings
will creep
        and fly
                and dance.

Typography by Kohar Alexanian
The text type is Gael, 14/18 composed by
The Haddon Craftsmen, Inc., Scranton, Penna.
The display type is Goudy Heavy Face, composed by
Cardinal Type Service, Inc.
Printed by Pearl Pressman Liberty, Printers
Bound by Economy Bookbinding Corp.
HARPER & ROW, PUBLISHERS, INCORPORATED